Magnets

Illustrations: Janet Moneymaker
Design/Editing: Marjie Bassler

Magnets
ISBN 978-1-953542-08-3

Published by Gravitas Publications Inc.
www.gravitaspublications.com
www.realscience4kids.com

REAL SCIENCE 4 Kids

Have you ever put a **magnet** on your refrigerator?

Have you ever pushed
a magnet through dirt?

Have you ever pushed one magnet close to another magnet?

What happens?

Magnets stick to surfaces,
pick up metal shavings,
push each other away, and
pull each other together.

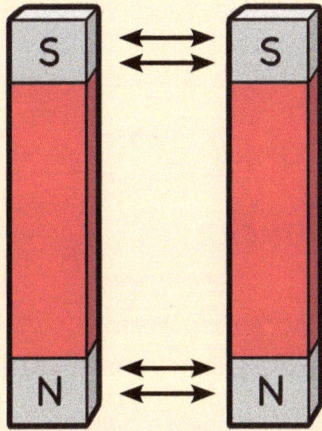

Magnets can push
each other away.

Magnets can pull
each other together.

But wait!
What is a
magnet?

Turn the page
to find out.

A magnet is made of metal **atoms** like **iron** or **nickel**.

Atoms are tiny building blocks that can link together.

Atoms make up everything we touch, taste, smell, and see.

The metal atoms line up like a box of marbles that are all pointing in the same direction. This makes a magnet have **poles**.

A magnet has two poles that are different from each other. One pole is at each end of the magnet.

I have a pole.

Not **that** kind of pole.

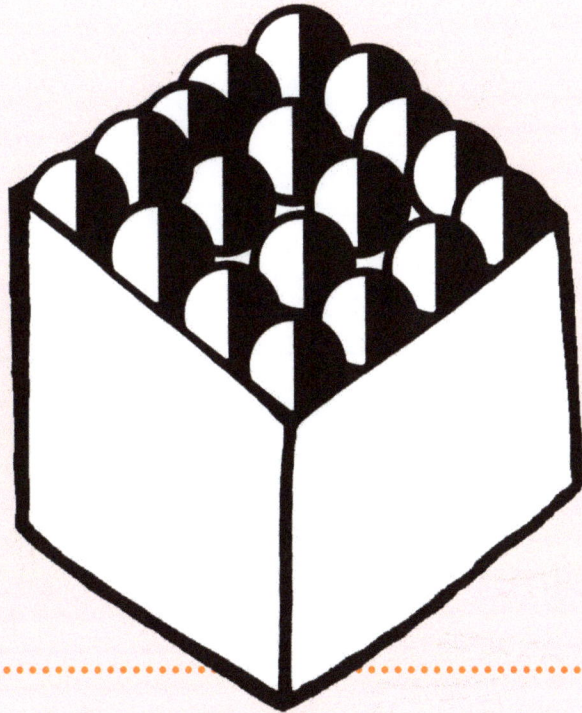

A **magnetic pole** creates a **force**.

In physics...

Force is any action that changes...

...the **location** of an object,

...the **shape** of an object,

...**how fast or how slowly** an object is

moving.

This force makes a magnet **attract**, or stick to other magnets and certain metals.

A **magnetic north pole** and a **magnetic south pole** have **opposite** forces. **Opposites attract.**

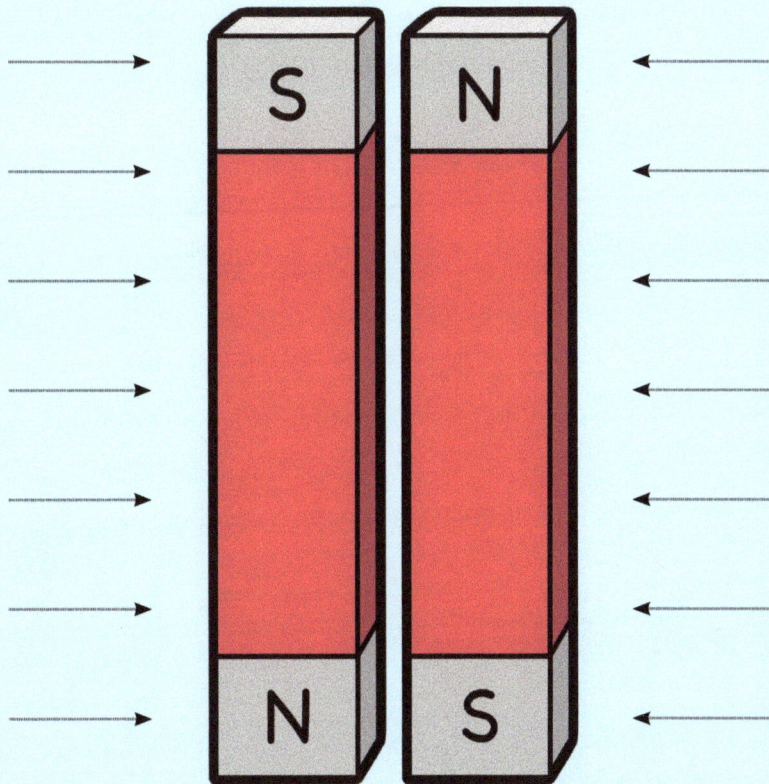

S N

N S

This force also makes a magnet **repel**, or push away, another magnet.

Magnetic poles that are
the same repel each other.

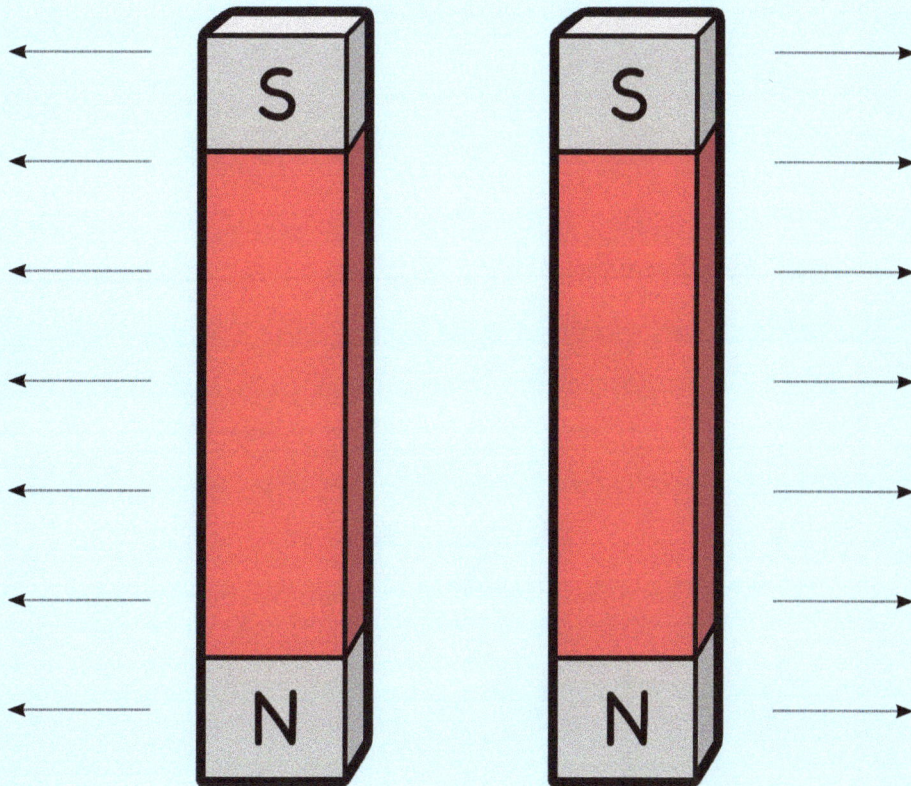

Magnets can attract and repel other magnets.

Magnets can attract some metals.

This means we can use magnets to...

How to say science words

atom   (AA-tum)

attract   (uh-TRAKT)

force   (FAWRSS)

iron   (IY-ern)

magnet   (MAG-net)

magnetic   (mag-NE-tik)

nickel   (NI-kuhl)

opposite   (AH-puh-zuht)

pole   (POHL)

repel   (rih-PEL)

What questions do you have about MAGNETS?

Learn More Real Science!

Complete science curricula from Real Science-4-Kids

Focus On Series

Unit study for elementary and middle school levels

Chemistry
Biology
Physics
Geology
Astronomy

Exploring Science Series

Graded series for levels K–8. Each book contains 4 chapters of:

Chemistry
Biology
Physics
Geology
Astronomy

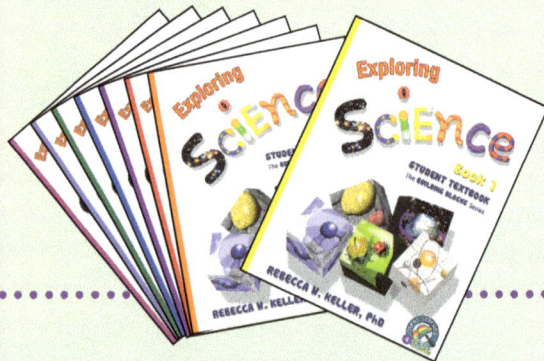

www.ingramcontent.com/pod-product-compliance
Lightning Source LLC
Chambersburg PA
CBHW040150200326
41520CB00028B/7556